How to prepare Welding Procedures for Oil & Gas Pipelines
according to API 1104 latest edition

Mohamed Ahmed Elsayed

P.E(Metallurgical and Materials)

This page intentionally left blank.

While every precaution has been taken in the preparation of this book, the publisher assumes no responsibility for errors or omissions, or for damages resulting from the use of the information contained herein.

How to prepare Welding Procedures for Oil & Gas Pipelines First edition. Apil, 2024.

Copyright © 2024 by Mohamed Ahmed Elsayed

All rights reserved.

This page intentionally left blank

Table of content

CHAPTER 1 SPECIFICATION INFORMATION -------------------- 1

CHAPTER 2 HOW TO PREPARE WPS MATRIX? -------------- 11

CHAPTER 3 MANUAL AND SEMIAUTOMATIC WELDING PROCESSES ESSENTIAL& NONESSENTIAL VARIABLES 13

CHAPTER 4 HOW TO FILL WPS MATRIX? ---------------------- 34

CHAPTER 5 HOW TO PREPARE PWPS? ------------------------ 37

pWPS Template --- 42

CHAPTER 6 HOW TO PREPARE PQR? -------------------------- 44

PQR Template-- 52

CHAPTER 7 HOW TO PREPARE WPS? ------------------------- 54

WPS Template --- 60

REFERENCES--- 63

v

This page intentionally left blank.

Introduction

API Standard 1104 – Welding of Pipelines and Related Facilities is the most widely-used industry standard in the world for pipeline construction. The use of this book will allow users to better understand how to prepare welding procedures for oil and gas pipelines according to API 1104 22nd edition. This book covers only manual and semiautomatic welding processes essential and nonessential variables and how to properly utilize them for welding procedures preparation.

This document includes the opinions of the author and not necessarily those of API.

This book results from many years spent in the fabricating and construction of products under the guidance and regulation of API 1104 standard, both national and international, that are used to help make pipeline systems safer.

This book is divided into seven chapters, which starts with specification information and its related mindmap, then how to prepare and fill WPS matrix, moving to understanding essential and nonessential variables, then the preparation of preliminary welding procedure (pWPS) & procedure qualification record (PQR), and finally welding procedure (WPS) preparation.

I have tried to include real examples to get the reader engaged as much as possible and provided a complete template at end of each chapter related to pWPS, PQR, and WPS for better understanding.

This page intentionally left blank.

Chapter 1
Specification Information

1.1. Scope:

In the scope we have to note that the API 1104 standard covers **carbon and low-alloy steel materials.** So, any other materials will be out of the scope of this standard.

Why do we select API 1104 for welding procedure qualification of pipelines and not ASME IX?

Because the welding qualification standard or code will be selected according to the requirements of the project construction code, for example: if the pipeline is for liquid transmission, so the construction code is ASME B31.4 in which it is mentioned that API 1104 shall be the welding qualification code.

1.2. What is meant by QUALIFIED WPS?

It is a tested and proven detailed method by which sound welds with suitable mechanical properties can be produced.

Which means that prior to start actual production a test coupon will be prepared (PQR) and we will apply all the proposed welding variables which we will use during actual production, such as type of welding process, base metal, filler metal, electrical characteristics, PWHT, etc. Then we will conduct mechanical tests (tensile test, bend test, nick break test, etc..) to check the weld soundness, and if it passes all the tests as per the API 1104 acceptance criteria, then it can be used in production as per the detailed WPS, which will be explained in the next chapters.

1.3. Qualification of Welding Procedures with Filler Metal Additions

Section 5 of API 1104 applies to the qualification of welding procedures using manual welding and semiautomatic welding processes with filler metal additions.

As per API 1104 the following specification information shall be included in the WPS, regardless of being essential or nonessential variables. Please refer to following mindmap

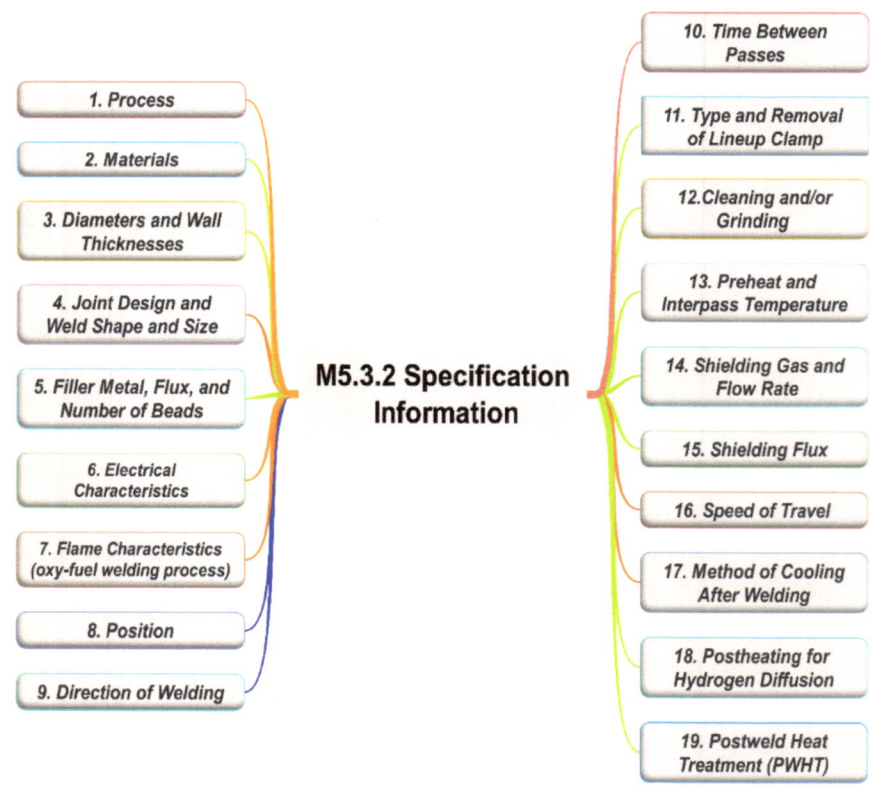

1.3.1. Process

The welding process and method of application shall be specified in the procedure either;
- Manual
- semiautomatic
- mechanized
- automatic
- combinations of these methods

For example:
Welding process: SMAW/FCAW
Method of application: Manual for SMAW/Semiautomatic for FCAW
SMAW for root pass, FCAW for hot pass, filling and capping

1.3.2. Materials

Procedure shall specify the applicable SMYS range
For example:
Base Metal qualified: API 5L Grade X65 and lower

1.3.3. Diameter and wall thicknesses

Despite, diameter is not essential variable and wall thickness is essential variable, but both of them shall be specified in the procedure
For example:
Outside Diameter: >12.750 in
thickness (t): 1.25 in. to unlimited

1.3.4. Joint Design and Weld Shape and Size

The procedure shall include a sketch or sketches of the joint, which shall show;
- tolerances for the angle of bevel

- the size of the root face
- the root opening or the space between abutting members
- the tolerance ranges for cap height and width for groove welds
- the shape and size tolerance ranges of fillet welds
- the type of backing if used

For example:
Joint Design: refer to following figure details

Backing: None
Joint type: V type butt joint
Root face(p): 1.6±0.8mm
Bevel angle (α): 30±2.5°
Root gap (b): 2.5-3.5mm

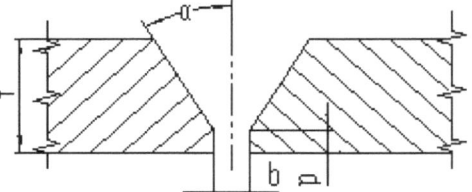

1.3.5. Filler Metal, Flux, and Number of Beads

The procedure shall specify;
- the sizes and classification of the filler metal and flux
- the minimum number and sequence of beads
- the number of beads and sequences for defined subsets of the thickness range

For example:

Weld Layer	Filler material		Polarity
	Type	Size (mm)	
Root (1)	E7010-P1	3.2	DCEN
Fill (2)	E8045-P2	4	DCEP
Cap (3)	E8045-P2	3.2	DCEP
Cap (4)	E8045-P2	3.2	DCEP

1.3.6. Electrical Characteristics

The procedure shall specify
- the current and polarity
- the range of voltage and amperage for each type and size of electrode, rod, or wire
- the heat input and the method of calculating heat input if required
    - for non-waveform-controlled processes; HI=60VA/(1000S)
    - for waveform-controlled processes; HI=E/L
    HI=PT/(1000L)

Where;
HI = heat input (kilojoules per in. or kilojoules per mm)
V = average welding arc voltage (volt)
A = average welding current (amp)
S = average travel speed (in. per minute or mm per minute)
E = average instantaneous energy measurement (kilojoules)
L = weld length (in. or mm)
P = average instantaneous power measurement (watt)
T = arc time (seconds)

For example:

Weld Layer	Process	Filler material		Polarity	Current (A)	Voltage (V)	Travel speed (mm/Min)	Heat input KJ/mm
		Type	Size (mm)					
Root	SMAW	E7010-P1	3.2	DCEN	90-120	22-28	80-110	1.49-1.83
Fill	SMAW	E8045-P2	4.0	DCEP	120-160	22-28	110 – 150	1.44-1.8
Cap	SMAW	E8045-P2	3.2	DCEP	90-120	22-28	80-110	1.49-1.83

1.3.7. Flame Characteristics (oxy-fuel welding process)

The procedure shall specify;
- whether the flame is neutral, carburizing, or oxidizing
- the size of the orifice in the torch tip for each size of rod or wire

1.3.8. Position

The procedure shall specify;
- roll or fixed position welding

For example:
Pipe Position: Fixed Welding/5G

1.3.9. Direction of Welding

The procedure shall specify;
- vertical up, vertical down, or horizontal direction

For example:
Welding Direction: Uphill (Root)
                     vertical down (Hot/Fill/Cap)

1.3.10. Time Between Passes

When using EXX10 or EXX11 electrodes, the procedure shall specify;
- maximum time between the completion of the root bead and the start of the second bead.
- maximum time between the completion of the second bead and the start of the third bead.
- the time shall be expressed as minutes or hours

For example:
Time between passes: Root bead to 2nd bead 10 minutes,
                       2nd bead to 3rd bead 2 hours

- As you notice the time between passes for Root bead and 2nd bead shall be set to the minimum possible because the weld joint is supported by the root pass only, which increases the possibility of cracking.
- For 2nd bead and 3rd bead the time can be increased considering the site conditions, like changing the welding process which will require mobilization of equipment and welders and may cause a delay in the start of the 3rd bead, this why you can specify the time up to 2 to 3 hours and even more.

1.3.11. Type and Removal of Lineup Clamp

The procedure shall specify;
- whether the lineup clamp is internal or external, or no clamp is required.
- the minimum percentage of root bead completed before the clamp is released.

For example:
Type of Line-up Clamp used: Internal line-up clamp
The internal line-up clamp shall not be removed before the completion of root bead.

1.3.12. Cleaning and/or Grinding

The procedure shall specify either;
- power tools
- hand tools
- or both

For example:
Interpass Cleaning: Power brushing and/or grinder.

1.3.13. Preheat and Interpass Temperature

The procedure shall specify;
- The method of heating and minimum preheat temperature immediately prior to welding
- The maximum interpass temperature

For example:
Min. Pre-heating Temperature: 60°C
Max. Inter-pass Temperature: 260°C
Pre-heating Method: Electrical heating or flame heating

1.3.14. Shielding Gas and Flow Rate

The procedure shall specify shielding gas classification and the range of flow rates
For example:
for GTAW: Shielding Gas: Argon
            Flow rate: 8-15 L/min
            Composition: ≥99.997 %

1.3.15. Shielding Flux

The procedure shall specify the type of shielding flux
For example:
for SMAW: Shielding Flux: NA

1.3.16. Speed of Travel

The procedure shall specify
- range for speed of travel for each pass or grouping of passes
- should be representative of the speed of travel used during procedure qualification

For example:

Weld Layer	Process	Filler material		Polarity	Current (A)	Voltage (V)	Travel speed (mm/Min)	Heat input KJ/mm
		Type	Size (mm)					
Root	SMAW	E7010-P1	3.2	DCEN	90-120	22-28	80-110	1.49-1.83
Fill	SMAW	E8045-P2	4.0	DCEP	120-160	22-28	110 – 150	1.44-1.8
Cap	SMAW	E8045-P2	3.2	DCEP	90-120	22-28	80-110	1.49-1.83

1.3.17. Method of Cooling After Welding

Specification shall designate
- the type of cooling after welding
- the maximum weld temperature prior to deliberate cooling

For example:
Method of Cooling after Welding: No forced cooling

1.3.18. Postheating for Hydrogen Diffusion

What is Postheating for Hydrogen Diffusion?
- Post-heating refers to the maintenance of preheat after the weld has been completed. It is required when there is a higher risk of cold cracking to allow increased rates of hydrogen evolution from the weld to occur.
- The post-heat temperature may be the same as, or greater than, the original preheat temperature specified and shall be done immediately after welding and before the weld region cools to below the minimum preheat temperature.

The procedure shall specify if applicable, The minimum temperature and time at temperature range.

For example:
Postheating temperature: 250°C to 300°C for 2 hr. minimum

1.3.19. Postweld Heat Treatment (PWHT)

The procedure shall specify;
- method of application
- heating rate
- temperature range
- time at temperature
- cooling rate

For example:
low alloy steel: Temperature Range 593-649°C
Time Range 1 hr./25mm WT, 1 hr. minimum.
Heating Rate 222 °C/hr./25mm WT
Cooling Rate 278 °C/hr./25mm WT

Chapter 2
How to prepare WPS matrix?

2.1 What is WPS matrix?
WPS matrix is a template in which the project specific information is summarized, in order to optimize the number of required PQRs to the minimum.

2.2 How to prepare WPS matrix?
i. From the engineering drawings and bill of material collect the required information (i.e., material type, Diameter, thickness, service and the toughness requirements)
ii. Identify the applicable construction code (i.e. ASME B31.4, B31.8, etc.) to use it for toughness, preheat and PWHT requirements
iii. Group the similar material types, Diameter, thickness, service and the toughness requirements to match the qualification code essential variables. Refer to next page for sample WPS matrix.
iv. Fill the first three columns of the template with the information we collected (material, Dia. & thickness).
v. The fourth column related to welding process will be filled based on the project budget and schedule. The welding process can be selected as manual, semi-automatic, automatic, mechanized or combination of these processes.
For our case we will select manual welding process SMAW as the main process.
Important Tip:
If SMAW is selected as the main welding process, always prepare another PQR for GTAW process to be utilized for welding of small bore (2 inch and below) materials (e.g., weld boss, branches, weldolet, etc.).

vi. Finally, the filler metal will be selected with consideration to following points;
- ➢ The compatibility of the base materials and filler metals from the standpoint of metallurgical and mechanical properties.
- ➢ The market availability and cost of the filler metal. Do not select any filler metal to qualify the WPS, and during production the required quantity is not available in the market or it is high in cost.
- ➢ The skills and experience of your welders with the selected filler metals. For example, some welders are well experienced with downhill filler metals, but when it comes to uphill, they are unable to produce sound welds.

After filling the first section of the matrix we need to find the best combination of the essential variables to reduce the number of PQRs to the minimum possible. This can be achieved by understanding section 5.4 of API 1104, which will be explained in details in our next chapter, and implement what we learn to fill all the columns of the WPS matrix and select the ideal test specimen which will cover the widest range of qualifications.

WPS Matrix template

Base Material	Project diameter range "Inch"	Required Project thickness range "mm"	W. process	Filler Metal	Test specimen			Qualified range		pWPS No.
					Material	Outer Dia. Inch	Thick. mm	Outer Dia mm	Thick mm	
API 5L Gr. X65	20":40"	8:16	SMAW	E7010-P1 E8045-P2						
API 5L Gr. X60	30":36"	8:14	SMAW	E7010-P1 E8045-P2						
API 5L Gr. X52	40":46"	9:16	SMAW	E6010 E7018-1						
API 5L Gr. X42	10":20"	6:9	SMAW	E6010 E7018-1						

Chapter 3
Manual and semiautomatic welding processes Essential & Nonessential variables

3.1 What is the difference between essential and nonessential variables?

- ***Essential variables*** are conditions in which a change, as described in the specific variables, is considered to affect the mechanical properties (other than toughness) of the joint. Before using a procedure specification whose essential variables have been revised and fall outside their qualified range, the procedure specification must be requalified.
- ***Nonessential variables*** are conditions in which a change, as described in the specific variables, is not considered to affect the mechanical properties of the joint. A procedure specification may be editorially revised to change a nonessential variable to fall outside of its previously listed range, but does not require requalification of the procedure specification.

3.2 Essential variables

As per API 1104, Section 5.4.2 & table 1, the essential variables are separated into two different categories.

1. Category I (standard WPS) essential variables shall apply when specified hardness and/or toughness values are not required by the company.
2. Category II (Hardness and/or Toughness) essential variables shall apply when hardness and/or toughness requirements are specified by the company.

As per API 1104, the essential variables are 14 items, and in order to make them easier for memorizing, the following mind map is created. However, in the mind map I divided the base metal qualifications into 2 items, one for SMYS and the other for thickness, so the total essential variables are 15 items, which we will go through them one by one.

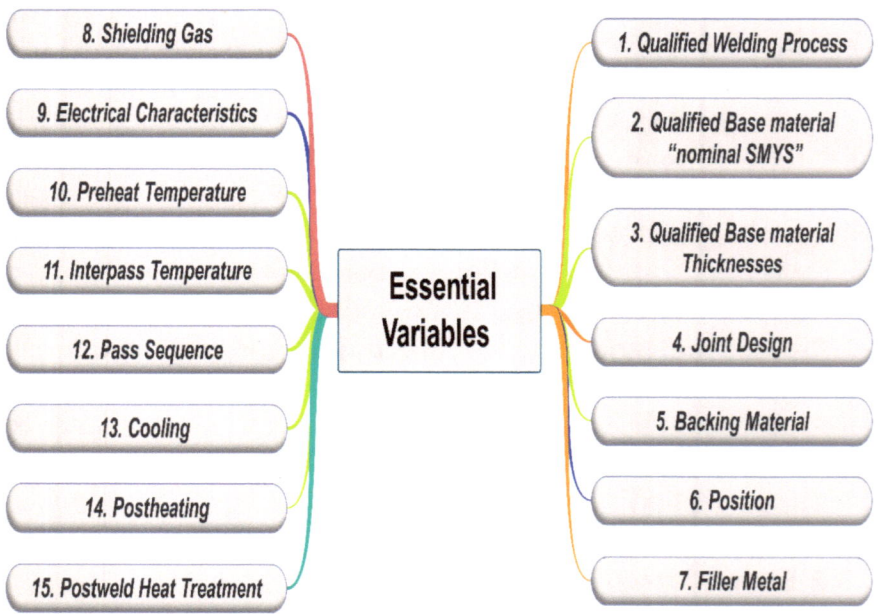

3.2.1 Qualified Welding Process

This item is divided into two points, which both of them are essential variables for both category I&II

3.2.1.1   A change in welding process for any given pass or pass grouping (i.e., fill passes or cap passes).

Which means for instance; if GTAW process is used for the root and SMAW process is used for hot, fill and cap passes, any change in this sequence, like utilizing GTAW for root and hot pass and the remaining passes with SMAW, will require requalification of the WPS.

3.2.1.2   A change between manual application or semiautomatic application.

Which means; if manual application used from the welding process as GTAW for example, a change to semiautomatic GTAW process requires requalification of the WPS.

3.2.2 Qualified Base material "nominal SMYS"

This item is divided into two points, which both of them are essential variables for both category I&II.

3.2.2.1   For same SMYS materials:
- The qualified SMYS equals to or less than the "nominal SMYS" used during qualification.

For example,
If the procedure was qualified using API5L Grade X65, then the qualified base metal is API5L Grade X65 and lower.

3.2.2.2   Two different SMYS materials
- At least one of the base materials is equal to or less than the lowest SMYS.

- The other material not greater than the maximum of the combination.

For example, if the PQR was done between two materials API5L Grade X65 and API5L Grade X52, so the qualified WPS range will be API 5L Grade X65 and lower to API 5L Grade X52 and lower.

Important tip:

If the PQR was qualified between the same two materials, it can still be used to weld two different materials during production, as long as both materials have nominal SMYS equal or less than the nominal SMYS of the material used during the qualification and all other essential variables are the same.

For example: **PQR:** *API5L Grade X65 to API5L Grade X65,*

Production *can weld API 5L Grade X60 to API5L Grade X52*

Key points for test coupon material selection:

A. Base material manufacturing process, heat treating process, carbon equivalent, and chemical composition should be considered for impact to mechanical properties and weld crack susceptibility;

For example,

- For a pipe material of a given grade, higher carbon equivalent materials generally have reduced weldability compared to lower carbon equivalent materials, so during PQR it is preferred to select the material with higher carbon equivalent, so if it passes the mechanical tests, it will give confidence that material with lower carbon equivalent will also pass.
- Same for heat treating process, for a pipe material of a given grade, a Quenched and tempered materials

generally have reduced weldability compared to Normalized and tempered materials, so during PQR it is preferred to select a quenched and tempered material so if it passes the mechanical tests, it will give confidence that Normalized and tempered materials will also pass.

B. When base material used during qualification has multiple grade markings, prior to qualification, the company should designate the material as one single grade.

For example,

- Prior to PQR if the material is marked to X56 and X52 the company shall designate the material as either X56 or X52
- Please also note that, as per API 5L multiple grade marking is allowed only for the following grade ranges:
  1) ≤ L290 (X42); can be marked together if it conforms to the requirements of each grade
  2) > L290 (X42) to < L415 (X60); can be marked together if it conforms to the requirements of each grade
  For instance;
  Manufacturer cannot mark the same pipe as X56 and X42 (different groups), but it can mark the pipe as X52 and X56, because both X56 and X52 are within the range of group 2.

3.2.3 Qualified Base material Thicknesses

This item is divided into two points, 1st point is essential variables for category I only and 2nd point is essential for category II only

3.2.3.1.1 **(For category I only)** where t is the nominal pipe wall thickness used in procedure qualification, a change that falls outside the following ranges:
   i. t to 2t when t < 0.154 in. (3.9 mm)

For example:
If the qualification thickness is **3 mm**, the qualified thickness will be **3 mm to 6 mm**

ii. 0.154 in. (3.9 mm) to 2t when 0.154 (3.9 mm) ≤ t ≤ 1.00 in. (25.4 mm).

For example:
If the qualification thickness is **15 mm**, the qualified thickness will be **3.9 mm to 30 mm.**

iii. 0.5t to unlimited when t > 1.00 in. (25.4 mm).

For example:
If the qualification thickness is **30 mm**, the qualified thickness will be **15 mm to unlimited.**

Important tip:

The PWHT requirements shall be verified as it will limit the qualified thickness.

For example: *if PWHT is required for material thickness **more than 32 mm as per project specifications** and the qualification thickness **without PWHT was 28 mm**, then the qualified base metal thickness will be **14 mm to 32 mm not to unlimited** as base metal thickness more than 32 mm requires PWHT and a new PQR is required to cover the thickness with PWHT.*

3.2.3.2 **(For category II only)** A change in nominal wall thickness with ± 25 % of the nominal thickness used during qualification.

For example:
If the qualification base metal thickness is 12 mm, so the 25% of 12 is 3, then the qualified base metal thickness will be 9 mm (12-3=9) to 15 mm (12+3=15)

3.2.4 *Joint Design*

This item is divided into two points, which both of them are essential variables for both category I&II

3.2.4.1   A change from fillet weld to groove weld, but not vice versa.

Which means that; if the qualification was carried out using fillet weld joint design, a change to groove weld requires requalification. However, if the qualification was done using groove weld joint design, a change to fillet weld does not require requalification.

3.2.4.2   A major change in joint type;

3.2.4.2.1   A change from the following major joint types (square, closed square, single-bevel, single-J, double-bevel, double-J, single-V, single-U, double-V, double-U) requires requalification.
For example:
a change from single bevel to single V requires requalification.

3.2.4.2.2   A change to or from a compound bevel within a major joint type ***does not require requalification***
For example:
a change of the bevel angle from 60° to 50° within the same joint type does not require requalification.

3.2.5 *Backing Material*

This item consists of only one point, which is essential variable for both category I&II

3.2.5.1   The deletion or change of backing material (steel, ceramic, non-ferrous, etc.)

This requirement applies to the root pass only. Weld metal is not considered weld backing.

For example:
If the qualification was done with ceramic as backing material, removing the backing or changing it to steel, for example, during the production requires requalification. However, if the qualification was done without backing material, the addition of backing during production does not require requalification.

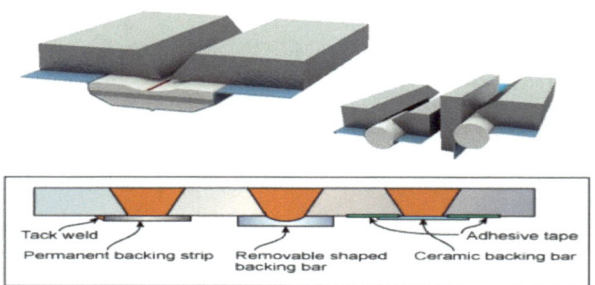

Remember that it is applicable for root pass only, i.e., if the backing material is removed during filling or cap passes, requalification is not required.

3.2.6 Position

This item consists of only one point, which is essential variable for both category I&II

3.2.6.1 A change in position from roll welding to fixed position, but not vice versa.

For example:
If the qualification was done by roll welding position, the WPS cannot be used for fixed positions and requalification is required. However, if the qualification was done by fixed welding position, the WPS can be used for roll welding position.

3.2.7 *Filler Metal*

This item consists of 10 points, 5 of them are essential variables for category I only, and three are essential variables for both category I & II, and the last two points are essential variables for category II only.

3.2.7.1   The first 5 points which are **applicable for category I** will be divided into two main groups;

3.2.7.1.1   For SMYS of **less than Grade X65**;

a) A change in filler metal grouping(s) as specified in Table 2 of API 1104

For example:
If the qualified base material is X60 and the filler metal is E8010-P1 (group 2) a change to E7010-P1 (group 1) requires requalification.

Table 2—Filler Metal Groups

Group	AWS Specification	AWS Classification
1	A5.1	E6010, E6011
1	A5.5	E7010-A1, E7010-P1, E7011-A1
2	A5.5	E8010-P1
3	A5.1	E6018, E7015, E7016, E7018, E7018M, E7016-1, E7018-1
3	A5.5	E7015-C1L, E7016-C1L, E7018-C1L, E7015-C2L, E7016-C2L, E7018-C2L, E7018-C3L, E8016-C1, E8018-C1, E8016-C2, E8018-C2, E8016-C3, E8018-C3, E8018-P2, E8045-P2
5 [a]	A5.18	ER70S-2, ER70S-3, ER70S-6
5 [a]	A5.28	ER70S-A1, ER80S-D2, ER80S-Ni1, ER80S-Ni2, ER80S-Ni3
6	A5.2	RG60, RG65
7	A5.20	E71T-1C, E71T-1M, E71T-9C, E71T-9M, E71T-12C, E71T-12M, E71T-12M-J
7	A5.36	E71T-1C, E71T-1M, E71T-9C, E71T-9M, E71T-12C, E71T-12M, E71T-12M-J
8	A5.29	E81T-1Ni1C, E81T-1Ni1M, E81T-1Ni2C, E81T-1Ni2M, E81T-1K2C, E81T-1K2M

[a] A shielding gas (see 5.4.2.7) is required for use with the electrodes in Group 5.

b) A change in sequence of deposition of filler metal groups when multiple filler metal groups are used for a single weld

For example:
if the qualified base material is X60 and the filler metal for root is E7010-P1 (group 1), and the fill& cap is E8018-C1 (group 3), a change in the sequence of deposition by doing the root and fill using group 1 and the cap by group 3 filler metal requires requalification.

c) For filler metals from group 1 through 3, any of the following changes require requalification;
- Any change in chemical composition designator except within the group, A1, C1, C2, C3, C1L, C2L, C3L, M, P1, P2
- A change in optional supplemental designator for toughness except addition of the optional supplemental designator for toughness (-1)
- A change in suffix designator from A1 to B3, or vice versa, constitutes an essential variable and requires requalification, but a change from A1 to C3, or vice versa, does not constitute an essential variable and does not require requalification.

For example:
A change from E7016 TO E7016-1, does not require requalification, however a change from E7016 to E7016-C1L requires requalification.

3.2.7.1.2 For SMYS of **Grade X65 and greater**;
a) A change in filler metal classification

For example:
If the qualified base material is X65 and the filler metal is E8016-C2, a change to E8018-C3 requires requalification, despite it is the same group and same AWS specification.

Please also note that the chemical composition designator is part of the AWS classification, so any change in this designator requires requalification

b) A change in sequence of deposition of filler metal classifications when multiple filler metal classifications are used for a single weld

For example:
If the qualified base material is X65 and the filler metal for root is E7016-C1L and the fill& cap is E8018-C1, a change in the sequence of deposition by doing the root and fill using E7016-C1L and only the cap by E8018-C1 filler metal requires requalification.

3.2.7.2　The following three points are **applicable for category I&II;**

3.2.7.2.1　A change in the manufacturer or trade name for filler metals with G designator only

For example:
If the procedure is qualified with E7018G from manufacturer (X), utilizing E7018G from manufacturer (Y) requires requalification. Also, if the procedure is qualified with E7018G from manufacturer (X), changing the trade name of the filler metal under the same manufacturer (X) requires requalification.

3.2.7.2.2　A change in AWS specification or AWS classification for filler metals not identified in Table 2.

3.2.7.2.3　A change in the nominal composition for filler metals not having an assigned AWS specification or classification.

3.2.7.3　The following two points are **applicable for category II only;**

3.2.7.3.1　A change in filler metal classification requires requalification, regardless of the base metal grade (X52, X60, X65, etc.).

3.2.7.3.2 A change in sequence of deposition of filler metal classifications when multiple filler metal classifications are used for a single weld requires requalification, regardless of the base metal grade (X52, X60, X65, etc.).

3.2.8 Shielding Gas

This item is divided into three points, which all of them are essential variables for both category I&II.

3.2.8.1 A change in shielding gas classification in accordance with AWS A5.32.
The shielding gases are classified as per table 4 of AWS A5.32 (Specification for Welding Shielding Gases)

**Table 4
AWS Classifications for
Typical Gas Mixtures**

AWS Classification	Typical Gas Mixtures (%)	Gas
SG-AC-25	75/25	Argon + Carbon Dioxide
SG-AO-2	98/2	Argon + Oxygen
SG-AHe-10	90/10	Argon + Helium
SG-AH-5	95/5	Argon + Hydrogen
SG-HeA-25	75/25	Helium + Argon
SG-HeAC-7.5/2.5	90/7.5/2.5	Helium + Argon + Carbon Dioxide
SG-ACO-8/2	90/8/2	Argon + Carbon Dioxide + Oxygen
SG-A-G	Special	Argon + Mixture

For example:
If the qualification used SG-AO-2 (98% Argon+ 2% Oxygen), a change to SG-AC-25 (75% Argon + 25% Carbon Dioxide) requires requalification.

3.2.8.2 Decrease in shielding gas flow rate by more than 20 % of that recorded during PQR.

For example:
If the nominal flow rate used during qualification is 20 L/min, a change to a nominal flow rate below 16 L/min requires requalification.
Why 16 L/min?
20% of 20L/min is 4 L/min, so 20-4= 16 L/min, which shall be the minimum flow rate that does not require requalification.

3.2.8.3   The deletion of or change in nominal composition of backing gas when backing gas is used during qualification.
For example:
If the qualification used backing gas, as in case of stainless-steel welding, deletion of the backing gas or changing the nominal composition requires requalification.

3.2.9 Electrical Characteristics

This item is divided into three points, which two of them are essential variables for both category I&II, and one is essential variable for category II only.

3.2.9.1   The following two points are **applicable for category I&II;**

3.2.9.1.1   A change in current/polarity type (DCSP/DCEN, DCRP/DCEP, AC).
The selection of type of current depends on application and welding process selected.
For example:
For GTAW welding process, DCEN is selected, as tungsten electrode is non-consumable and the arc heat is required to be concentrated on the base metal not the consumable, but in the case of SMAW process

DCEP is mostly selected as the arc heat need to be concentrated to melt the SMAW electrode. Therefore, changing current/polarity requires requalification.

3.2.9.1.2 A change to or from a waveform-controlled process.

What is waveform-controlled process?

- ❖ It is the ability of a welding power source to affect heat input, droplet shape and size, penetration, bead shape and toe wetting by the use of microprocessor controls which manage the welding output.
  - ➢ This normally a feature used in semi-automatic & automatic application, but it is also available in some manual welding machines.

  For example:
  If the qualification was done with a waveform-controlled process, a change to non- waveform-controlled process requires requalification and vice versa.

3.2.9.2 The following point is applicable **for category II only**;

3.2.9.2.1 A change in heat input exceeding ± 20 % of that recorded during qualification requires requalification.
The heat input is calculated as per section 5.3.2.6 of API 1104, and recorded as averages representative of a weld pass or specific segment of a weld pass.

For example:
If the recorded average heat input during qualification is 1 kj/mm, the qualified range shall be 0.8 to 1.2 kj/mm.

3.2.10 Preheat Temperature

This item is divided into two points, which both of them are essential variables for both category I&II

3.2.10.1   Any decrease in minimum base material preheat temperature below the base material preheat temperature recorded during qualification requires requalification.

For example:
If the base metal preheat temperature during qualification is 60°C, any decrease below this temperature requires requalification.

3.2.10.2   If preheat is not required, the minimum temperature specified shall be the **lesser of**:
- 60°F (16°C) or
- the actual base material temperature recorded prior to qualification welding.

For example:
In case preheat is not required, and during qualification the base metal temperature recorded prior to welding was 20°C, then the WPS shall specify the minimum preheat temperature as 16°C.

3.2.11 Interpass Temperature

This item is divided into two points, 3.2.11.1 is essential variables for category I&II and 3.2.11.2 is essential for category II only

3.2.11.1   An increase in base material interpass temperature to greater than 500°F (260°C).

This means that the maximum interpass temperature for all materials under the scope of API 1104 is 260°C. So, during qualification you shall ensure that the interpass temperature measured immediately prior to the start of subsequent weld passes is not exceeding 260°C.

3.2.11.2 An increase in base material interpass temperature greater than 100°F (55°C) above the maximum base material interpass temperature recorded during procedure qualification.

For example:
If the maximum interpass temperature recorded during qualification was 150°C, then the maximum qualified interpass temperature is 150+55= 205°C

Important tip:
Please take into consideration point 3.2.11.1 related to maximum interpass temperature of 260°C, while implementing point 3.2.11.2
For example:
If the maximum interpass temperature recorded during qualification was 230°C, then the maximum qualified interpass temperature is 230+55= 285°C, which is higher than the 260°C requirements. Hence the qualified maximum interpass temperature in this case shall be specified in the WPS as 260°C instead of 285°C.

3.2.12 Pass Sequence

This item consists of only one point, which is essential variable for both category I&II

3.2.12.1 A change in bead deposition sequence, when using a temper bead technique requires requalification

For example,
a change from a deposition sequence that relies on tempering to some other deposition sequence requires requalification.

What is temper bead welding?
It is a weld bead placed at a specific location in or at the surface of a weld for the purpose of affecting the metallurgical properties of the heat-affected zone or previously deposited weld metal. See below examples of typical temper bead deposition sequences.

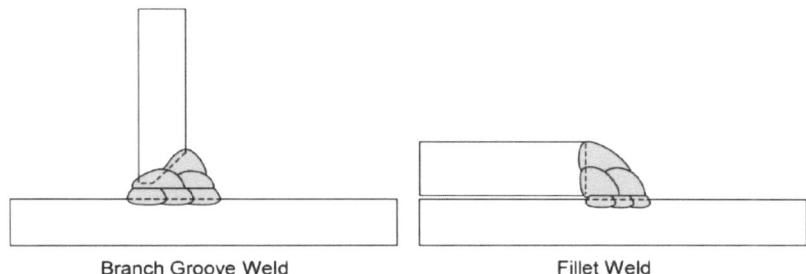

Branch Groove Weld   Fillet Weld

NOTES:
1. A layer of weld metal buttering is first deposited using stringer beads.
2. Higher heat input levels are used for subsequent passes, which refine and temper the HAZ of the first layer.

3.2.13 Cooling

This item is divided into three points, which all of them are essential variables for category I&II.

3.2.13.1 The addition or deletion of deliberate cooling methods requires requalification.

For example:
If forced cooling with water was applied during qualification, the deletion of it in production requires requalification.

3.2.13.2 A change in the method of deliberate cooling after welding, resulting in a higher rate of cooling requires requalification.

For example:
If the weld joint was cooled during qualification under controlled rates (For example: 100°C/hr.), a change to forced cooling with water during production will

produce higher cooling rates, which will require requalification.

3.2.13.3 An increase in the maximum weld temperature prior to deliberate cooling requires requalification.

For example:
If the deliberate cooling was recorded during qualification to start at max temperature of 400°C, then any increase in temperature above 400°C prior the deliberate cooling requires requalification.

3.2.14 Postheating

This item is divided into three points, which all of them are essential variables for category I&II.

3.2.14.1 The elimination of postheating for the purpose of promoting hydrogen diffusion.

As explained earlier postheating is required when there is a higher risk of cold cracking to allow increased rates of hydrogen release from the weld to occur.
Therefore, if postheating was recorded during qualification, the elimination of it requires requalification

3.2.14.2 A reduction in postheating temperature greater than 60 °F (33 °C) from that used during qualification.

For example:
If the postheating temperature recorded during qualification was 300°C, the minimum qualified postheating shall be 300-33=267°C

3.2.14.3 Any reduction in postheating time at temperature from that used during qualification.

For example:
If the postheating time at 300°C was 2 hr., any postheating time at the same temperature for less than 2 hr. requires requalification.

3.2.15 Postweld Heat Treatment

This item is divided into two points, which both of them are essential variables for category I&II

3.2.15.1 The addition or deletion of PWHT requires requalification.
If the procedure was qualified without PWHT, the addition of PWHT requires requalification and vice versa.
3.2.15.2 When applied, a change in PWHT procedure requires requalification.

> **For example:**
> Changes in the PWHT method of application, heating rate, temperature range, time at temperature, and cooling rate, out of the qualified range in the WPS, requires requalification

3.3 Non-Essential Variables

After reviewing the essential variables and compare them with the specification information in section 5.3.2 of API 1104, we can conclude the non-essential variables as following;

3.3.1 Diameters

Despite diameter is non-essential variable, the ranges of specified outside diameters (ODs) over which the procedure is applicable shall be specified.

For example:
Outside Diameter range: ≥6 inch

3.3.2 Direction of Welding

The specification shall indicate whether the welding is to be performed in the vertical up, vertical down, or horizontal direction.

For example:
Welding Direction: Vertical up (Root)
                    Vertical down (Fill/Cap)

3.3.3 Time Between Passes

When using EXX10 or EXX11 electrodes, the maximum time between the completion of the root bead and the start of the second bead, as well as the maximum time between the completion of the second bead and the start of the third bead, shall be specified.

For example:
Time between passes: Root bead to 2nd bead **10 minutes,**
                       2nd bead to 3rd bead **2 hours**

3.3.4 Type and Removal of Lineup Clamp

The specification shall indicate whether the lineup clamp is to be internal or external, or if no clamp is required. If a clamp is used, the minimum percentage of root bead completed before the clamp is released shall be specified

For example:
Type of Line-up Clamp:
- Internal clamp for pipe diameter16 inches or larger. For pipe diameter less than 16 inch, either internal or external line-up clamps may be used.
- The internal line-up clamp shall not be removed before the completion of the root bead. For external clamps, the root bead must be at least 50% complete prior to removal.

3.3.5 Cleaning and/or Grinding

The specification shall indicate whether power tools or hand tools are to be used for cleaning, grinding, or both.

For example:
Interpass Cleaning: Power brushing and/or grinder.

3.3.6 Voltage, Amperage, Filler metal size & Travel Speed

The changes in voltage and amperage from the ranges used during qualification are non-essential variables, however the filler metal manufacturer's recommended ranges shall be considered for each filler metal sizes.

Note: In case of hardness or toughness requirements, we shall take into consideration that any changes in voltage, amperage and travel speed from the one recoded during qualification, shall not cause a change in heat input exceeding ± 20% of that recorded during qualification.

Chapter 4
How to fill WPS matrix?

After we got a clear understanding of essential and non-essential variables from the previous chapter, it is the time to fill our WPS matrix to plan the required number of procedure qualifications.

Below is the matrix template that was prepared in previous chapters, so let's start filling all the missing information step by step.

Before starting, please note that we will discuss, base material, diameter, thickness, welding process, and filler metals, as the main variables of matrix preparation, as the other items can be selected by the company and not a project requirement, except for PWHT, which is rarely used in pipelines.

Base Material	Project diameter range "Inch"	Required Project thickness range "mm"	W. process	Filler Metal	Test specimen			Qualified range		pWPS No.
					Material	Outer Dia. Inch	Thick. mm	Outer Dia mm	Thick mm	
API 5L Gr. X65	20":40"	8:16	SMAW	E7010-P1 E8045-P2						
API 5L Gr. X60	30":36"	8:14	SMAW	E7010-P1 E8045-P2						
API 5L Gr. X52	40":46"	9:16	SMAW	E6010 E7018-1						
API 5L Gr. X42	10":20"	6:9	SMAW	E6010 E7018-1						

4.1 Base Material

As explained in previous chapter the material with the highest nominal SMYS qualifies all the lower nominal SMYS base materials. Hence from the matrix we need to select the highest nominal SMYS, which is API 5L Gr. X65. However, the filler metal classifications are different for X65&X60(E7010-P1&

E8045-P2) and X52&X42(E6010& E7018-1) which is essential variable for category II WPS as explained earlier, therefore two test specimen materials (X65 & X52) need to be selected.

4.2 Diameter

Diameter is not an essential variable; hence any diameter could be selected to cover all the ranges in the project. So, we can select 10 inches for our qualification.

4.3 Thickness

From the matrix the thickness range is 6-16 mm, as per table 2, 5.4.2.2 (b & c), there are two scenarios:

a- For Category I (standard WPS), we can select 10 mm as the qualification coupon thickness. So, the qualified range is 3.9 mm to 20 mm, which will cover the entire project range.

b- Category II (Hardness and/or Toughness), one thickness cannot cover the entire range, so we shall select two specimens at least to cover the entire project thickness range;

- The first specimen thickness selected is 8 mm, which will cover a qualified range (25% less than 8mm) 6mm to (25% greater than 8mm) 10 mm.
- The second specimen thickness selected is 13 mm, which will cover a qualified range (25% less than 13 mm) 9.75 mm to (25% greater than 13mm) 16.25 mm.

As we can see from the above thickness selections for category II, the entire project range is covered (6 mm to 16 mm).

Important tip:

We can use the same PQR for category II as supporting to category I WPS, and apply the requirements of category I WPS only.

From the previous steps we can adjust the matrix to reflect the selected specimens, in order to prepare the required pWPSs, as following;

Final WPS matrix

Base Material	Project diameter range "Inch"	Required Project thickness range "mm"	W. process	Filler Metal	Test specimen			Qualified range		pWPS No.
					Material	Outer Dia. Inch	Thick. mm	Outer Dia mm	Thick mm	
API 5L Gr. X65	20":40"	8:16	SMAW	E7010-P1 E8045-P2	API 5L Gr. X65	10 inches	8 mm	≥2.375 inch	6-10 mm	pWPS-01
API 5L Gr. X60	30":36"	8:14	SMAW	E7010-P1 E8045-P2			13 mm		9.75-16.25 mm	pWPS-02
API 5L Gr. X52	40":46"	9:16	SMAW	E6010 E7018-1	API 5L Gr. X52	10 inches	8 mm	≥2.375 inch	6-10 mm	pWPS-03
API 5L Gr. X42	10":20"	6:9	SMAW	E6010 E7018-1			13 mm		9.75-16.25 mm	pWPS-04

In the next chapter we will prepare one of the pWPSs, utilizing the WPS matrix data.

Chapter 5
How to prepare pWPS?

In this chapter we will learn how to prepare pWPS, step by step;

But first, we need to know what is pWPS?

pWPS is Preliminary or Proposed Welding Procedure Specification, in which the proposed variables to be applied during production are introduced.

These variables are required to be validated prior implementing them in production, through the Procedure Qualification to check the weld soundness, and if it passes all the tests as per the API 1104 acceptance criteria, then the pWPS will be transferred to qualified WPS and can be used in production.

Steps of pWPS preparation

API 1104 figure 2, provides a sample WPS, which we will modify it to adapt the changes required to be included in the pWPS.

Let's start

5.1 General section

The first section of pWPS is general, which includes the company name, pWPS No/ PQR, date, applicable code, etc. and can be filled as following;

Company Name: ABC	Date: 1-June-2023
Preliminary Welding Procedure Specification No.: PWPS-01	Rev. No.: 0
Supporting PQR No.: PQR-01C.S	
Welding Process: SMAW Type: Manual	Process Application: Main Pipeline Welding
Reference Standard: API 1104 22nd edition	Other: ASME B31.4 & ASME B31.8

5.2 Base Material

In order to fill the information in this section, we will use the WPS matrix to find the proposed test coupon material Grade, Outside Diameter and Wall Thickness, as following;

Base Material	
Test Coupon Material Grade	API 5L Gr.X65 to API 5L Gr.X65
Test Coupon Dia. & Thickness	Dia. 10", Thickness 8mm
Outside Diameter Range	≥ 2.375" (60.3 mm)
Thickness Range	6-10 mm

5.3 Joint Design

The joint design can be decided based on agreement between production/construction team and quality team, and this section can be filled as shown below;

Joint Design	
Joint Design: See detail figure	Joint type: single-V, Groove weld
Groove angle (α) : 30±2.5°	Root face (p) : 1.6±0.8 mm
Root gap (b) : 2.5-3.5mm	Cap bead width: 0.5-2.0 mm from each edge of groove
Backing: N/A	Reinforcement: 0.8-2.0 mm

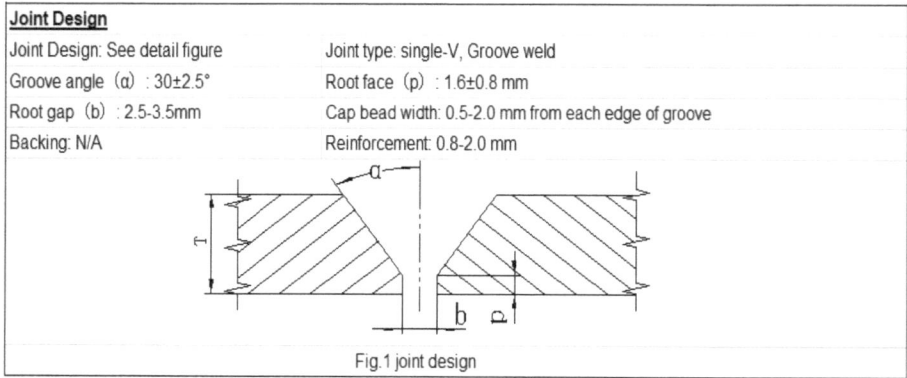

Fig.1 joint design

5.4 Filler Metal

In order to fill the information in this section, we will use the proposed filler metals in the WPS matrix as following;

Filler Metal		
Weld Layer	Root	Fill & Cap
Filler Metal Group No.	1	3
AWS Spec.& Classification	AWS A5.5 E7010-P1	AWS A5.5 E8045-P2
Size	Ø 3.2mm	Ø 3.2, 4.0mm
Welding Direction	Downhill	Downhill

5.5 POSITIONS

It is preferred to select the pipe position as fixed position, as it is not practical to roll the pipe during pipeline welding at site.
This part can be filled as following;

POSITIONS
Pipe Position: Fixed position

5.6 Shielding Gas

The welding process proposed is SMAW, hence shielding GAS is not applicable, and it can be filled as following;

GAS	
Shielding Gas:	NA
Composition:	NA
Flow Rate:	NA
Gas Backing:	NA

5.7 Preheat & Interpass Temperature

The preheat temperature selection depends on the construction code and/or project specification, and it is selected according to base material thickness. For our case we will consider that no Preheat is required, however it shall be expressed in the pWPS as per API 1104 table 1.

For the interpass temperature, the maximum is 260°C, as explained in previous chapters

Both preheat temperature & interpass temperature can be expressed in the pWPS as shown below;

PREHEAT
Min. Pre-heating temperature: the lesser of 60 °F (16 °C) or the actual base material temperature recorded prior to qualification welding
Max. Inter-pass Temperature: 260°c
Pre-heating Method: Electrical heating or flame heating

5.8 Postheating & Method of cooling

This section can be filled as following;

POSTHEATING	METHOD OF COOLING
Postheating NA	Method of cooling: No forced cooling

5.9 Postweld Heat Treatment (PWHT)

The PWHT selection, depends on the construction code and/or project specification, and it is selected according to base material thickness and/or service. For our case we will consider that no PWHT is required.

It can be expressed in the pWPS as shown below;

POSTWELD HEAT TREATMENT
PWHT: NA

5.10   Technique

TECHNIQUE REQUIREMENTS
Number of Welders: 2 welders operating simultaneously for each pass
Time between passes: Root bead to 2nd bead 10 minutes, 2nd bead to 3rd bead 2 hours
String or Weave: String / Weave
Orifice or Gas Cup Size(mm): NA
Method of Back Gouging: NA
Multiple or Single Electrodes: Single
Initial Inter-Pass Cleaning: Brushing and/or grinding. Etc.
Multiple or Single Pass (per side): Multiple
Electrode drying: according to manufacturer's requirement
Type of Line-up Clamp used: Internal or external Clamp
• For pipelines less than 16-inch diameter, either internal or external line-up clamps may be used. • Internal or External line-up clamp shall be used if the pipe diameter is 16 inches or larger, 100% RT to be done if used external clamp. a) The internal line-up clamp shall not be removed before the completion of the root bead. b) external clamps, the root bead must be at least 50% complete prior to removal

5.11 ELECTRICAL CHARACTERISTICS

All related information can be addressed in the following table

Welding parameters

Weld Layer	Process	Filler material		Polarity	Current (A)	Voltage (V)	Travel speed (mm/Min)	Heat input KJ/mm
		Type	Size (mm)					
Root	SMAW	E7010-P1	3.2	DCEN	90-120	22-28	80-110	1.49-1.83
Fill	SMAW	E8045-P2	4	DCEP	120-160	22-28	110 – 150	1.44-1.8
Cap	SMAW	E8045-P2	3.2	DCEP	90-120	22-28	80-110	1.49-1.83

How to prepare Welding Procedures for Oil & Gas Pipelines

pWPS Template

Company Name: ABC		**Date:** 1-June-2023	
Preliminary Welding Procedure Specification No.: PWPS-01		**Rev. No.:** 0	
Supporting PQR No.: PQR-01C.S			
Welding Process: SMAW **Type:** Manual		**Process Application:** Main Pipeline Welding	
Reference Standard: API 1104 22rd edition		**Other:** ASME B31.4 & ASME B31.8	
Base Material			
Test Coupon Material Grade	API 5L Gr.X65 to API 5L Gr.X65		
Test Coupon Dia. & Thickness	Dia. 10", Thickness 8mm		
Outside Diameter Range	≥ 2.375" (60.3 mm)		
Thickness Range	6–10 mm		
Joint Design			
Joint Design: See detail figure	Joint type: single-V, Groove weld		
Groove angle（α）: 30±2.5°	Root face (p) : 1.6±0.8 mm		
Root gap（b）: 2.5–3.5mm	Cap bead width: 0.5–2.0 mm from each edge of groove		
Backing: N/A	Reinforcement: 0.8–2.0 mm		
Fig.1 joint design			
Filler Metal			
Weld Layer	Root		Fill & Cap
Filler Metal Group No.	1		3
AWS Spec.& Classification	AWS A5.5 E7010-P1		AWS A5.5 E8045-P2
Size	Ø 3.2mm		Ø 3.2, 4.0mm
Welding Direction	Downhill		Downhill
POSITIONS		**POSTWELD HEAT TREATMENT**	
Pipe Position: Fixed position		PWHT:	NA
PREHEAT		**GAS**	
Min. Pre-heating temperature: the lesser of 60 °F (16 °C) or the actual base material temperature recorded prior to qualification welding		Shielding Gas:	NA
Max. Inter-pass Temperature: 260°C		Composition:	NA
Pre-heating Method: Electrical heating or flame heating		Flow Rate:	NA
		Gas Backing:	NA

How to prepare Welding Procedures for Oil & Gas Pipelines

POSTHEATING		METHOD OF COOLING	
Postheating NA		Method of cooling: No forced cooling	
TECHNIQUE REQUIREMENT			
Number of Welder: 2 welders operating simultaneously for each pass			
Time between passes: Root bead to 2nd bead 10 minutes, 2nd bead to 3rd bead 2 hours			
String or Weave: String / Weave		**Initial Inter-Pass Cleaning:** Brushing and/or grinding. Etc.	
Orifice or Gas Cup Size(mm): NA		**Contact Tube to Work Dist.(mm):** NA	
Method of Back Gouging: NA		**Multiple or Single Pass (per side):** Multiple	
Multiple or Single Electrodes: Single		**Electrode drying:** according to manufacturer's requirement	
Type of Line-up Clamp used: Internal or external Clam			
For pipelines less than 16-inch diameter, either internal or external line-up clamps may be used.			
Internal or External line-up clamp shall be used if the pipe diameter is 16 inches or larger, 100% RT to be done if used external clamp.			
a) The internal line-up clamp shall not be removed before the completion of the root bead.			
b) external clamps, the root bead must be at least 50% complete prior to removal			
ELECTRICAL CHARACTERISTICS			
Current (AC or DC): See table below		Polarity: See table below	
Amps (Range): See table below		Volts (Range): See table below	
Travel speed: See table below		Tungsten Electrode type & size: NA	
Welding parameters			

Weld Layer	Process	Filler material		Polarity	Current	Voltage	Travel speed	Heat input
		Type	Size (mm)		(A)	(V)	(mm/Min)	KJ/mm
Root	SMAW	E7010-P1	3.2	DCEN	90-120	22-28	80-110	1.49-1.83
Fill	SMAW	E8045-P2	4	DCEP	120-160	22-28	110 – 150	1.44-1.8
Cap	SMAW	E8045-P2	3.2	DCEP	90-120	22-28	80-110	1.49-1.83

Approved By:

Chapter 6
How to prepare PQR?

In order to prepare the PQR we need to divide the process into three stages;
1- Stage 1 (Prior to Welding of test coupon)
2- Stage 2 (during Welding of test coupon)
3- Stage 3 (After Welding of test coupon)

In the following section, each stage will be discussed in details to get clear understanding of how the PQR is prepared for any kind of welding processes.

6.1 Stage 1 (Prior to Welding of test coupon)

a) Verify that the test coupon is as per pWPS (material type, size, dimension, etc.)

b) Consider the required number of mechanical tests in your selection of the test coupon dimension. For pipe less than 2.375 in. (60.3 mm) in OD, two test joints shall be made to obtain the required number of test specimens.

c) The minimum number of test specimens and the tests to which they shall be subjected are given in Table 3.

Table 3—Type and Number of Test Specimens for Procedure Qualification Test

Outside Diameter of Pipe		Number of Specimens					
in.	mm	Tensile Strength	Nick Break	Root Bend	Face Bend	Side Bend	Total
Wall Thickness ≤ 0.500 in. (12.7 mm)							
< 2.375	< 60.3	0 [a]	2 [b]	2	0	0	4 [c]
2.375 to 4.500	60.3 to 114.3	0 [a]	2 [b]	2	0	0	4
> 4.500 to 12.750	>114.3 to 323.9	2	2 [b]	2	2	0	8
>12.750	> 323.9	4	4 [b]	4	4	0	16
Wall Thickness > 0.500 in. (12.7 mm)							
≤ 4.500	≤ 114.3	0 [b]	2 [b]	0	0	2	4
> 4.500 to 12.750	> 114.3 to 323.9	2	2 [b]	0	0	4	8
> 12.750	> 323.9	4	4 [b]	0	0	8	16

[a] For materials with nominal SMYS greater than Grade X42, a minimum of one tensile test is required.
[b] Nick break tests are not required for procedure qualification butt welds, provided procedure welds are examined by radiography or ultrasonic testing (UT) and are found acceptable per Section 9. When radiography or UT results are unacceptable, the procedure is rejected.
[c] One nick break and one root bend specimen are taken from each of two test welds, or for pipe less than or equal to 1.315 in. (33.4 mm) in diameter, one full-section tensile strength specimen is take.

d) Select experienced welders to weld the PQR test coupon
e) Fix the coupon on the required position as per the pWPS (roll, fixed, 5G or 6G etc.), as shown in the following figure

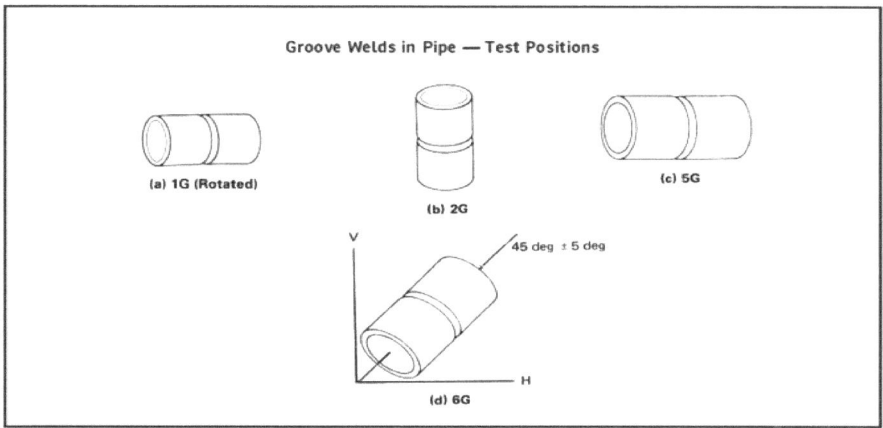

f) Verify the welding consumables as per pWPS and follow manufacturer's recommendations for Conditioning, Storage, and Exposure.
g) Ensure that all the equipment utilized are calibrated as applicable

6.2 Stage 2 (during Welding of test coupon)
a) Record all the essential variables during the PQR
b) Record the supplementary essential variables when toughness test is required as per the applicable construction code
c) Non-essential variables may be recorded based on the organization option
d) All the recorded variables shall be the actual values

Let's start the PQR form filling in order to get clear understanding.

6.2.1 General section

The first section of PQR is general, which includes the company name, WPS No/ PQR, date, applicable code, etc. and can be filled as following;

Company Name: ABC	Date: 1-June-2023
Welding Procedure Specification No.: WPS-01	Rev. No.: 0
Supporting PQR No.: PQR-01C.S	Rev. No.: 0
Welding Process: SMAW Type: Manual	Process Application: Main Pipeline Welding
Reference Standard: API 1104 22nd edition	Other: ASME B31.4 & ASME B31.8

6.2.2 Base material and Joint design

In this section, the PQR includes the base material data, dia., thickness & joint design used during the actual PQR, and can be filled as following;

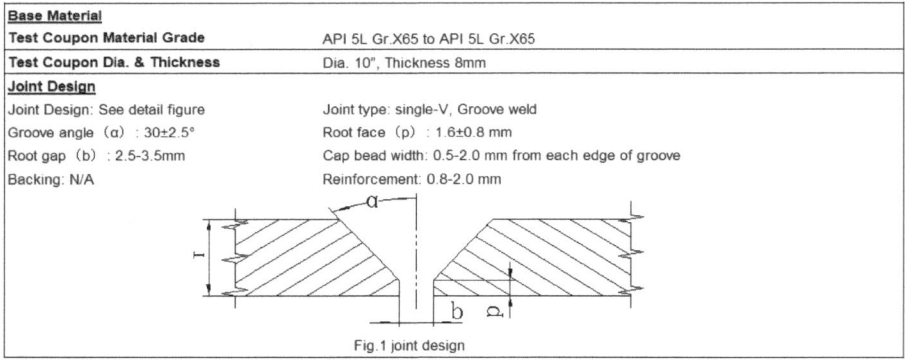

Fig.1 joint design

6.2.3 Filler Metal

In this section the PQR includes the filler metal requirements which were utilized during welding of the test coupon including filler metal group, Classification, diameter etc. and can be filled as following;

Filler Metal		
Weld Layer	Root	Fill & Cap
Filler Metal Group No.	1	3
AWS Spec.& Classification	AWS A5.5 E7010-P1	AWS A5.5 E8045-P2
Size	Ø 3.2mm	Ø 3.2, 4.0mm
Welding Direction	Downhill	Downhill
Deposited weld thickness	3 mm	5 mm

6.2.4 POSITIONS
Pipe Position: Fixed position

6.2.5 Shielding Gas
The welding process proposed is SMAW, hence shielding GAS is not applicable, and it can be filled as following

Shielding Gas: NA

6.2.6 Preheat:

PREHEAT	GAS
Min. Pre-heating temperature: 20°C	Shielding Gas: NA
Max. Inter-pass Temperature: refer to welding parameters table	Composition: NA
Pre-heating Method: Electrical heating or flame heating	Flow Rate: NA
	Gas Backing: NA

6.2.7 Postheating and Method of cooling

POSTHEATING	METHOD OF COOLING
Postheating NA	Method of cooling: No forced cooling

6.2.8 Postweld Heat Treatment (PWHT)
No PWHT, and It can be expressed in the PQR as shown below;

PWHT: NA

6.2.9 Technique

TECHNIQUE REQUIREMENTS
Number of Welders: 2 welders operating simultaneously for each pass
Time between passes: Root bead to 2nd bead 10 minutes, 2nd bead to 3rd bead 2 hours
String or Weave: String / Weave
Method of Back Gouging: NA
Multiple or Single Electrodes: Single
Initial Inter-Pass Cleaning: Brushing and/or grinding. Etc.
Multiple or Single Pass (per side): Multiple
Electrode drying: according to manufacturer's requirement
Type of Line-up Clamp used: external Clamp. It was removed after completion of 50% of the root pass

6.2.10 Recorded Welding Parameters

The welding parameters can be recorded in the PQR as following;

Note: Amperage and voltage are the minimum and maximum recorded during qualification.

Welding parameters								
Weld Layer	Filler material		Polarity	Interpass Temperature °C	Current	Voltage	Travel speed	Heat input
	Type	Size (mm)			(A)	(V)	(mm/Min)	KJ/mm
Root (1)	E7010-P1	3.2	DCEN	20	90-100	22-27	90 - 110	1.32 – 1.5
Fill (2)	E8045-P2	4	DCEP	230	120-150	23 - 28	130 - 150	1.3 – 1.68
Cap (3)	E8045-P2	3.2	DCEP	190	90-110	24 - 28	105 - 125	1.23 – 1.5
Cap (4)	E8045-P2	3.2	DCEP	200	100-120	24 - 28	100 - 115	1.44 – 1.75

6.3 Stage 3 (After Welding of test coupon)

a) Perform NDT to ensure weld soundness prior to mechanical test. Radiography or ultrasonic testing (UT) shall be carried out and interpreted as per API 1104, Section 9.
b) Perform mechanical test as specified in API 1104, Para 5.6.
c) Reflect the mechanical test results in any format fits your organization.
d) The PQR shall be certified accurate by the Organization

6.3.1 Applicable Mechanical tests

- As the test coupon size is 10" Dia. & 8mm thickness, so from table 3 the number of specimens can be selected as highlighted below

Table 3—Type and Number of Test Specimens for Procedure Qualification Test

Outside Diameter of Pipe		Number of Specimens					
in.	mm	Tensile Strength	Nick Break	Root Bend	Face Bend	Side Bend	Total
Wall Thickness ≤ 0.500 in. (12.7 mm)							
< 2.375	< 60.3	0 a	2 b	2	0	0	4 c
2.375 to 4.500	60.3 to 114.3	0 a	2 b	2	0	0	4
> 4.500 to 12.750	>114.3 to 323.9	2	2 b	2	2	0	8
>12.750	> 323.9	4	4 b	4	4	0	16
Wall Thickness > 0.500 in. (12.7 mm)							
≤ 4.500	≤ 114.3	0 b	2 b	0	0	2	4
> 4.500 to 12.750	> 114.3 to 323.9	2	2 b	0	0	4	8
> 12.750	> 323.9	4	4 b	0	0	8	16

a For materials with nominal SMYS greater than Grade X42, a minimum of one tensile test is required.
b Nick break tests are not required for procedure qualification butt welds, provided procedure welds are examined by radiography or ultrasonic testing (UT) and are found acceptable per Section 9. When radiography or UT results are unacceptable, the procedure is rejected.
c One nick break and one root bend specimen are taken from each of two test welds, or for pipe less than or equal to 1.315 in. (33.4 mm) in diameter, one full-section tensile strength specimen is take.

- Location of Test Butt Weld Specimens for Procedure Qualification Test can be found as per **figure 3 of API 1104,** as shown below

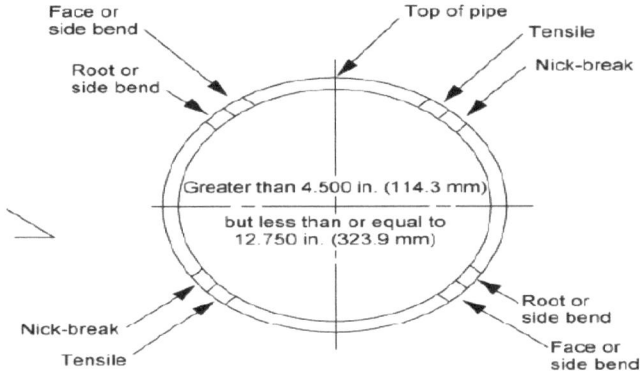

6.3.1.1 The Tensile Strength Test Evaluation

The tensile strength test can be evaluated as following;
   a) The tensile strength of the weld shall be greater than or equal to the specified minimum tensile strength (SMTS) of the pipe material, but need not be greater than or equal to the actual tensile strength of the material.
   b) If the specimen breaks outside the weld (i.e., in the parent metal) at a tensile strength not less than 95 % of that of the SMTS of the pipe material, the weld shall be accepted as meeting the requirements.
   c) If the specimen breaks in the weld and the observed strength is greater than or equal to the SMTS of the pipe material and meets the soundness requirements of **API 1104, Para 5.6.3.3**, the weld shall be accepted as meeting the requirements.

The requirements of API 1104, Para 5.6.3.3 can be summarized as following;
   I. specimen shall show complete penetration and fusion.
   II. The greatest dimension of any gas pocket shall not exceed 1/16 in. (1.6 mm), and the combined area of all

gas pockets shall not exceed 2 % of the exposed surface area.

III. Slag inclusions shall not be more than 1/32 in. (0.8 mm) in depth and shall not be more than 1/8 in. (3 mm) or one-half the specified wall thickness in length, whichever is smaller. There shall be at least ½ in. (13 mm) separation between adjacent slag inclusions of any size.

IV. The dimensions should be measured as shown in Figure 6. Fisheyes, as defined in AWS A3.0, are not cause for rejection.

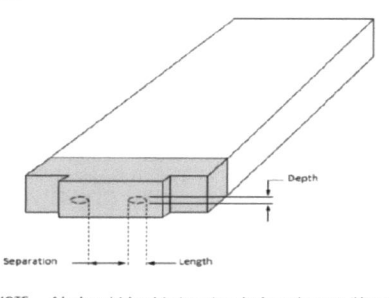

NOTE A broken nick break test specimen is shown; however, this method of dimensioning applies also to broken tensile and fillet weld test specimens.

Figure 6—Dimensioning of Imperfections in Exposed Weld Surfaces

6.3.1.2   The Nick Break Test Evaluation

Nick break tests are not required for procedure qualification butt welds, provided procedure welds are examined by radiography or ultrasonic testing (UT) and are found acceptable per Section 9 of API 1104. When radiography or UT results are unacceptable, the procedure is rejected.

The nick break test can be evaluated as following;

a) specimen shall show complete penetration and fusion.

b) The greatest dimension of any gas pocket shall not exceed 1/16 in. (1.6 mm), and the combined area of all gas pockets shall not exceed 2 % of the exposed surface area.

c) Slag inclusions shall not be more than 1/32 in. (0.8 mm) in depth and shall not be more than 1/8 in. (3 mm) or one-

half the specified wall thickness in length, whichever is smaller. There shall be at least ½ in. (13 mm) separation between adjacent slag inclusions of any size.
d) The dimensions should be measured as shown in Figure 6 of API 1104. Fisheyes, as defined in AWS A3.0, are not cause for rejection.
e) For a test weld diameter greater than 12 ¾ in. (323.9 mm), if only one nick break specimen fails, the company shall have the discretion to either consider the weld unacceptable or to replace the specimen with two additional nick break specimens from locations as close as possible to the failed specimen. If either of the replacement nick break specimens fail, the weld shall be considered unacceptable.

6.3.1.3 The Bend Test Evaluation

The Bend test can be evaluated as following;
a) The bend test shall be considered acceptable if no crack or other imperfection exceeding 1/8 in. (3 mm) or one-half the specified wall thickness, whichever is smaller, in any direction is present in the weld or between the weld and the fusion zone after bending.
b) Cracks that originate on the outer radius of the bend along the edges of the specimen during testing and that are less than ¼ in. (6 mm), measured in any direction, shall not be considered unless obvious imperfections are observed.
c) For a test weld diameter greater than 12 ¾ in. (323.9 mm), if only one bend specimen fails, the company shall have the discretion to consider the weld unacceptable or to replace the failed specimen with two additional specimens from locations adjacent to the failed specimen. If either of the replacement bend test specimens fails, the weld shall be considered unacceptable.

How to prepare Welding Procedures for Oil & Gas Pipelines

PQR Template

Company Name: ABC		Date: 1-June-2023	
Welding Procedure Specification No.: WPS-01		Rev. No.: 0	
Supporting PQR No.: PQR-01C.S		Rev. No.: 0	
Welding Process: SMAW Type: Manual		Process Application: Main Pipeline Welding	
Reference Standard: API 1104 22nd edition		Other: ASME B31.4 & ASME B31.8	
Base Material			
Test Coupon Material Grade	API 5L Gr.X65 to API 5L Gr.X65		
Test Coupon Dia. & Thickness	Dia. 10", Thickness 8mm		
Joint Design			
Joint Design: See detail figure	Joint type: single-V, Groove weld		
Groove angle (α) : 30±2.5°	Root face (p) : 1.6±0.8 mm		
Root gap (b) : 2.5-3.5mm	Cap bead width: 0.5-2.0 mm from each edge of groove		
Backing: N/A	Reinforcement: 0.8-2.0 mm		
Fig.1 joint design			
Filler Metal			
Weld Layer	Root		Fill & Cap
Filler Metal Group No.	1		3
AWS Spec.& Classification	AWS A5.5 E7010-P1		AWS A5.5 E8045-P2
Size	Ø 3.2mm		Ø 3.2, 4.0mm
Welding Direction	Downhill		Downhill
Deposited weld thickness	3 mm		5 mm
POSITIONS		**POSTWELD HEAT TREATMENT**	
Pipe Position: Fixed position		PWHT: NA	
PREHEAT		**GAS**	
Min. Pre-heating temperature: 20°C		Shielding Gas: NA	
Max. Inter-pass Temperature: refer to welding parameters table		Composition: NA	
Pre-heating Method: Electrical heating or flame heating		Flow Rate: NA	
		Gas Backing: NA	

How to prepare Welding Procedures for Oil & Gas Pipelines

POSTHEATING	METHOD OF COOLING
Postheating NA	Method of cooling: No forced cooling

TECHNIQUE REQUIREMENT

Number of Welder: 2 welders operating simultaneously for each pass

Time between passes: Root bead to 2nd bead 10 minutes,
                          2nd bead to 3rd bead 2 hours

String or Weave: String / Weave	**Initial Inter-Pass Cleaning:** Brushing and/or grinding. Etc.
Method of Back Gouging: NA	**Multiple or Single Pass (per side):** Multiple
Multiple or Single Electrodes: Single	**Electrode drying:** according to manufacturer's requirement
Type of Line-up Clamp used: external Clamp	external clamp was removed after completion of 50% of the root pass

ELECTRICAL CHARACTERISTICS

Current (AC or DC): See table below	Polarity: See table below
Amps (Range): See table below	Volts (Range): See table below
Travel speed: See table below	Tungsten Electrode type & size: NA

Welding parameters

Weld Layer	Filler material		Polarity	Interpass Temperature °C	Current (A)	Voltage (V)	Travel speed (mm/Min)	Heat input KJ/mm
	Type	Size (mm)						
Root (1)	E7010-P1	3.2	DCEN	20	90-100	22-27	90 - 110	1.32 – 1.5
Fill (2)	E8045-P2	4	DCEP	230	120-150	23 - 28	130 - 150	1.3 – 1.68
Cap (3)	E8045-P2	3.2	DCEP	190	90-110	24 - 28	105 - 125	1.23 – 1.5
Cap (4)	E8045-P2	3.2	DCEP	200	100-120	24 - 28	100 - 115	1.44 – 1.75

Approved By:

Chapter 7
How to prepare WPS?

After the procedure qualification completion, the essential, non-essential and supplementary essential variables shall be reflected in the WPS as following;

7.1 General Section

The first section of PQR& WPS is General section, which includes the company name, PQR/WPS No, date applicable code, etc. and can be filled as following;

Company Name: ABC	Date: 1-July-2023
Welding Procedure Specification No.: WPS-01	Rev. No.: 0
Supporting PQR No.: PQR-01C.S	Rev. No.: 0
Welding Process: SMAW Type: Manual	Process Application: Main Pipeline Welding
Reference Standard: API 1104 22nd edition	Other: ASME B31.4& ASME B31.8

7.2 Base Material

- Base material is **essential variable**. The material used during PQR is API 5L Gr.X65, hence the qualified material is API 5L Gr.X65 and lower as explained earlier in Chapter 3.
- Outside Diameter is non-essential variable, so the procedure can be qualified for all diameters, however it is not preferred to utilize SMAW for welding the root of small diameter (≤2").
- Pipe wall thickness is essential variable. As the procedure is qualified with toughness, so the qualified base metal thickness is *6-10 mm* (plus or minus 25 % of 8mm).

The Base material, Outside Diameter& Pipe wall thickness can be reflected in the WPS as following;

Base Material	
Test Coupon Material Grade	API 5L Gr.X65 and lower to API 5L Gr.X65 and lower
Outside Diameter Range	≥ 2.375" (60.3 mm)
Thickness Range	6-10 mm

7.3 Joint Design

- Major change in joint type is essential variable, however changes in bevel angle within the same joint type is not a reason of requalification.
- The PQR is qualified for single-V, Groove weld, so joint design can be expressed in the WPS as following;

Joint Design	
Joint Design: See detail figure	Joint type: single-V, Groove weld
Groove angle (α) : 30±2.5°	Root face (p) : 1.6±0.8 mm
Root gap (b) : 2.5-3.5mm	Cap bead width: 0.5-2.0 mm from each edge of groove
Backing: N/A	Reinforcement: 0.8-2.0 mm

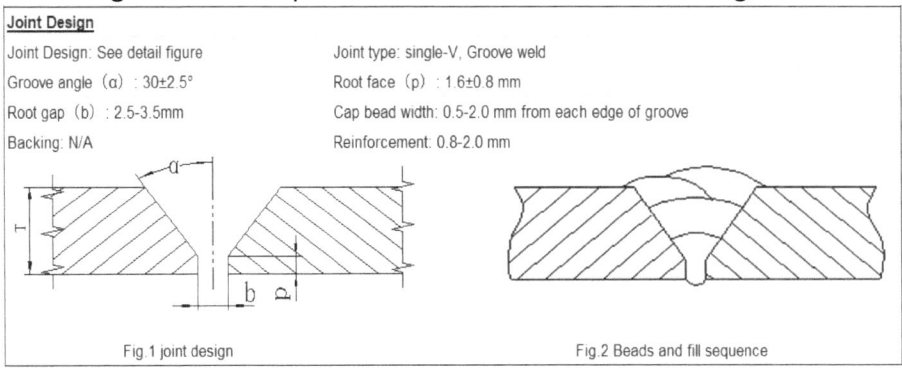

Fig.1 joint design   Fig.2 Beads and fill sequence

7.4 Filler Metal

- As the procedure is qualified with toughness, hence the WPS is qualified for the same filler metal classification and same sequence of deposition of filler metal classifications used in the PQR.
- Size of filler metal is not essential variable; however, the heat input shall be controlled within the qualified range.
- Direction of Welding is not essential variable, however filler metal manufacturer's recommendations should be followed

The WPS can be filled as following;

Filler Metal		
Weld Layer	Root	Fill & Cap
Filler Metal Group No.	1	3
AWS Spec.& Classification	AWS A5.5 E7010-P1	AWS A5.5 E8045-P2
Size	Ø 3.2mm	Ø 3.2, 4.0mm
Welding Direction	Downhill	Downhill
Max. Thickness of Deposited Weld Metal	2.25 - 3.75 mm	3.75 - 6.25 mm

Please note that the qualified thickness of deposited weld metal for each filler metal shall be reflected in the WPS.

7.5 Position

Position is essential variable and as the PQR is carried out at fixed position, hence the WPS is qualified for Fixed and roll positions.
The WPS can be filled as following;

Pipe Position: Fixed and roll positions

7.6 Shielding Gas

The welding process is SMAW, hence shielding GAS is not applicable, and WPS can be filled as following

Shielding Gas: NA

7.7 Postweld Heat Treatment (PWHT)

PWHT is essential variable and as the procedure was qualified without PWHT, so the WPS shall reflect the same as shown below

PWHT: NA

7.8 Preheat & Interpass temperature

- Preheat is essential variable and as the Preheat was not applied to the qualification test weld, hence the minimum temperature reflected in the WPS shall be the lesser of 60 °F (16 °C) or the actual base material temperature recorded prior to qualification welding (20°C). Hence the selected Preheat temperature is **16 °C.**
- The interpass temperature is essential variable and it shall be reflected for each pass, considering that it shall not increase by greater than 55 °C above the maximum base

material interpass temperature recorded during procedure qualification and in all cases, it shall not exceed 260°C.

It can be clarified as following;

Weld Layer	Recorded Interpass Temperature	Qualified Interpass Temperature
Root	20(preheat Temp.)	20(preheat Temp.)
Fill	230	230+55=285, but the temperature shall not exceed 260, so the WPS shall specify 260
Cap	190	190+55=245 the WPS can specify 245

The minimum preheat temperature & method and the maximum interpass temperature can be specified in the WPS as following;

PREHEAT

Min. Pre-heating temperature: 16 °C

Max. Inter-pass Temperature: 16°C for root, 260°C for fill & 245°C for cap

Pre-heating Method: Electrical heating or flame heating

7.9 Postheating

Eliminating Postheating is essential variable, and as the PQR is qualified without Postheating, hence WPS can reflect the same, as illustrated below

Postheating: NA

7.10 Method of cooling

Method of cooling is essential variable and it shall be specified in the WPS as recorded in the PQR.

WPS can be specified as following;

Method of cooling: No forced cooling

7.11 Technique

Time between passes, Number of Welders, Cleaning and/or grinding and Type & Removal of Lineup Clamp are non-essential variables which can be specified in the WPS as following;

TECHNIQUE REQUIREMENTS
Number of Welders: 2 welders operating simultaneously for each pass
Time between passes: Root bead to 2nd bead 10 minutes, 2nd bead to 3rd bead 2 hours
String or Weave: String / Weave
Orifice or Gas Cup Size(mm): NA
Method of Back Gouging: NA
Multiple or Single Electrodes: Single
Initial Inter-Pass Cleaning: Brushing and/or grinding. Etc.
Multiple or Single Pass (per side): Multiple
Electrode drying: according to manufacturer's requirement
Type of Line-up Clamp used: Internal or external Clamp
• For pipelines less than 16-inch diameter, either internal or external line-up clamps may be used. • Internal or External line-up clamp shall be used if the pipe diameter is 16 inches or larger, 100% RT to be done if used external clamp. a) The internal line-up clamp shall not be removed before the completion of the root bead. b) external clamps, the root bead must be at least 50% complete prior to removal

7.12 Welding parameters

- Welding parameters can be reflected in the WPS as following

Welding parameters

Weld Layer	Electrical characteristics	Filler material		Polarity	Current	Voltage	Travel speed	Heat input
		Type	Size (mm)		(A)	(V)	(mm/Min)	KJ/mm
Root	non-waveform-controlled	E7010-P1	3.2	DCEN	90-120	22-28	90 - 120	1.32 - 1.68
Fill	non-waveform-controlled	E8045-P2	4	DCEP	120-160	22-28	130 - 150	1.22 - 1.8
Cap	non-waveform-controlled	E8045-P2	3.2	DCEP	90-120	23-28	100 - 115	1.24 - 1.75

- A change in current/polarity type is essential variable, so they shall be reflected in the table as per PQR (DCEN for root & DCEP for fill and cap).
- A change to or from a waveform-controlled process is essential variable, so the table shall reflect it as per the PQR (non-waveform-controlled).
- The changes in voltage and amperage from the ranges used during qualification is not essential variables, however the filler metal manufacturer's recommended ranges shall be considered.
- Heat input shall be calculated without a change of more than ±20 % of that recorded during qualification.

7.13 Approval

By completing the above sections, our WPS is almost completed and only pending one final step, which is approving the WPS for production.
This can be reflected in the WPS as following

Approved By:

How to prepare Welding Procedures for Oil & Gas Pipelines

WPS Template

Company Name: ABC		Date: 1-July-2023	
Welding Procedure Specification No.: WPS-01		Rev. No.: 0	
Supporting PQR No.: PQR-01C.S		Rev. No.: 0	
Welding Process: SMAW	Type: Manual	Process Application: Main Pipeline Welding	
Reference Standard: API 1104 22nd edition		Other: ASME B31.4 & ASME B31.8	
Base Material			
Test Coupon Material Grade	API 5L Gr.X65 and lower to API 5L Gr.X65 and lower		
Outside Diameter Range	≥ 2.375" (60.3 mm)		
Thickness Range	6-10 mm		
Joint Design			
Joint Design: See detail figure	Joint type: single-V, Groove weld		
Groove angle (α) : 30±2.5°	Root face (p) : 1.6±0.8 mm		
Root gap (b) : 2.5-3.5mm	Cap bead width: 0.5-2.0 mm from each edge of groove		
Backing: N/A	Reinforcement: 0.8-2.0 mm		

Fig.1 joint design Fig.2 Beads and fill sequence

Filler Metal		
Weld Layer	Root	Fill & Cap
Filler Metal Group No.	1	3
AWS Spec.& Classification	AWS A5.5 E7010-P1	AWS A5.5 E8045-P2
Size	Ø 3.2mm	Ø 3.2, 4.0mm
Welding Direction	Downhill	Downhill
Max. Thickness of Deposited Weld Metal	2.25 - 3.75 mm	3.75 - 6.25 mm

POSITIONS		**POSTWELD HEAT TREATMENT**	
Pipe Position: Fixed and roll positions		PWHT:	NA
PREHEAT		**GAS**	
Min. Pre-heating temperature: 16 °C		Shielding Gas:	NA
Max. Inter-pass Temperature: 16°c for root, 260°c for fill & 245°c for cap		Composition:	NA
Pre-heating Method: Electrical heating or flame heating		Flow Rate:	NA
		Gas Backing:	NA

How to prepare Welding Procedures for Oil & Gas Pipelines

POSTHEATING	METHOD OF COOLING
Postheating NA	Method of cooling: No forced cooling

TECHNIQUE REQUIREMENT

Number of Welder: 2 welders operating simultaneously for each pass

Time between passes: Root bead to 2nd bead 10 minutes,
2nd bead to 3rd bead 2 hours

String or Weave: String / Weave   **Initial Inter-Pass Cleaning:** Brushing and/or grinding. Etc.

Method of Back Gouging: NA   **Multiple or Single Pass (per side):** Multiple

Multiple or Single Electrodes: Single   **Electrode drying:** according to manufacturer's requirement

Type of Line-up Clamp used: Internal or external Clamp

For pipelines less than 16-inch diameter, either internal or external line-up clamps may be used.

Internal or External line-up clamp shall be used if the pipe diameter is 16 inches or larger, 100% RT to be done if used external clamp.

a) The internal line-up clamp shall not be removed before the completion of the root bead.

b) external clamps, the root bead must be at least 50% complete prior to removal

ELECTRICAL CHARACTERISTICS

Current (AC or DC): See table below   Polarity: See table below

Amps (Range): See table below   Volts (Range): See table below

Travel speed: See table below   Tungsten Electrode type & size: NA

Welding parameters

Weld Layer	Electrical characteristics	Filler material		Polarity	Current (A)	Voltage (V)	Travel speed (mm/Min)	Heat input KJ/mm
		Type	Size (mm)					
Root	non-waveform-controlled	E7010-P1	3.2	DCEN	90-120	22-28	90 - 120	1.32 - 1.68
Fill	non-waveform-controlled	E8045-P2	4	DCEP	120-160	22-28	130 - 150	1.22 - 1.8
Cap	non-waveform-controlled	E8045-P2	3.2	DCEP	90-120	23-28	100 - 115	1.24 - 1.75

Approved By:

This page intentionally left blank.

References

1- *API STANDARD 1104 22nd EDITION, WELDING OF PIPELINES AND RELATED FACILITIES.*
2- *API Specification 5L, Specification for Line Pipe*
3- *AWS A3.0, Standard Welding Terms and Definitions*
4- *AWS A5.32 Specification for Welding Shielding Gases*

ABOUT THE AUTHOR

Mohamed Ahmed Elsayed is P.E(Metallurgical and Materials) with extensive sixteen (16) years of work experience in the field of Quality Assurance, Quality Control, Welding, Quality Inspections, etc. Mr. Mohamed holds several well recognized certifications, including ASQ CMQ/OE, ASNT Level III (RT, UT, PT&MT), API 510, API 570, CWI and more.

Mr. Mohamed is a co-author of two books: one being a mindmap companion book for the ASQ Certified Manager of Quality – Organizational Excellence, and the other is Welding Procedure Specification (WPS) & Procedure Qualification Record (PQR) Practical Guide.

This page intentionally left blank.

www.ingramcontent.com/pod-product-compliance
Lightning Source LLC
Chambersburg PA
CBHW040229220526
45473CB00001B/171